by Mary Dylewski

Orlando Austin New York San Diego Toronto London

Visit *The Learning Site!*
www.harcourtschool.com

Going to the Market

It is Saturday. Lila is going to the farmer's market with her dad. She grabs the list they made and runs to the car. She jumps into the back seat and puts on her safety belt. Her dad smiles at her in the mirror. "Let's try to classify the things we will buy today," he says to Lila.

"What does classify mean?" asks Lila.

"It means to put things into groups," her dad says. "You can classify things by the ways they are the same."

A farmer's market

At the market, Lila looks around at all the stands. She sees bananas, apples, pears, and oranges. "How can we classify these things?" Lila asks.

"How are they the same?" her dad asks.

"They are all fruits," says Lila. "They all have peels. They taste sweet."

"Great job, Lila," says her dad. "Let's classify some other things here. Find some things that have the same shape."

◀ **You can buy many kinds of fruits and vegetables at a farmer's market.**

Shape and Texture

Lila looks around. "Tomatoes and apples are round," she says. "So are these onions. The peppers and bananas are long and thin."

"Very good, Lila," says her dad. He pointed to another fruit stand. "See if you can classify these fruits by texture."

"What is texture?" asks Lila.

"Texture is the way something feels. An apple feels smooth. A lemon feels bumpy. A pineapple is rough and prickly."

Which of these fruits have the same shape? ▼

You can classify fruits by their textures. ▼

"Bananas and apples are smooth," says Lila. "Oranges and limes feel bumpy. This coconut is rough. The little kiwis feel fuzzy!"

Smell and Taste

"That's right, Lila," says her dad. "Now, how else could you classify things here?"

"Hmmm," says Lila. She looks around.

"Close your eyes," says her dad. "Take a deep breath. Can you classify things now?"

"Yes," says Lila. "The peaches and melons smell sweet. The mushrooms smell like soil when it rains!"

◀ You can classify fruits by their taste.

"Now let's use our sense of taste to classify some more things," says Lila's dad.

"Yum!" says Lila. "I'm hungry!"

Lila and her dad buy and taste some fruits. The purple plums taste very sweet. The red ones are sour. The grapefruits are sour, too, but the strawberries taste sweet.

"Now I know another way to classify things by texture," said Lila.

"How?" asks her dad.

"By eating them!" says Lila. "Bananas are soft. Coconut is hard. A mango is stringy. Tiny seeds make strawberries crunchy!"

Color

"That's great, Lila!" says her dad. "Let's classify one more way before we leave. Can you classify things by color?"

"That's easy," says Lila. "These apples and strawberries are red. The bananas and lemons are yellow. These grapes and plums are dark purple. The pears over there are red and yellow and green. Let's go buy some. Pears are my favorite fruit!"

You can also classify fruits by their color. ▼

Flexibility

Lila takes a slice of red pear. She can bend it and it doesn't break. She takes a slice of green pear. She can't bend it. Her dad explains that things that can bend are flexible. Flexibility is another way to classify things.

Lila and her dad buy the things on their list. They put their bags into the car. On the way home, they talk about other ways to classify fruits and vegetables. Lila thinks of classifying them by size. Her dad thinks of classifying them by weight.

How would you classify them?

◀ **You can classify fruits and vegetables in many ways.**